LET'S CODE!

CODING CONDITIONALS

BY DONNA McKINNEY

CONTENT CONSULTANT
DAVID A. LASH
ASSOCIATE PROFESSOR OF COMPUTER SCIENCE
AURORA UNIVERSITY

Kids Core
An Imprint of Abdo Publishing
abdobooks.com

abdobooks.com

Printed in the United States of America, North Mankato, Minnesota.
102023
012024

THIS BOOK CONTAINS
RECYCLED MATERIALS

Cover Photo: Olivier Le Moal/Shutterstock Images
Interior Photos: Shutterstock Images, 4–5, 7, 10, 14, 20, 26, 28 (bottom), 29 (top); Wavebreak Media/Shutterstock Images, 9; Ground Picture/Shutterstock Images, 12–13; Ahmad Denny Syahputra/Shutterstock Images, 15; Carlos Amarillo/Shutterstock Images, 16; Amado Designs/Shutterstock Images, 19 (clocks); Red Line Editorial, 19 (chart); Irina Wilhauk/Shutterstock Images, 22–23, 28 (top); Monkey Business Images/Shutterstock Images, 25, 29 (bottom)

Editor: Katharine Hale
Series Designer: Katharine Hale

Library of Congress Control Number: 2023939666

Publisher's Cataloging-in-Publication Data

Names: McKinney, Donna B., author.
Title: Coding conditionals / by Donna B. McKinney
Description: Minneapolis, Minnesota: Abdo Publishing, 2024 | Series: Let's code! | Includes online resources and index.
Identifiers: ISBN 9781098292751 (lib. bdg.) | ISBN 9798384910695 (ebook)
Subjects: LCSH: Coding theory--Juvenile literature. | Computer programming--Juvenile literature. | Computers and children--Juvenile literature. | Conditionals--Juvenile literature.
Classification: DDC 005.1--dc23

CONTENTS

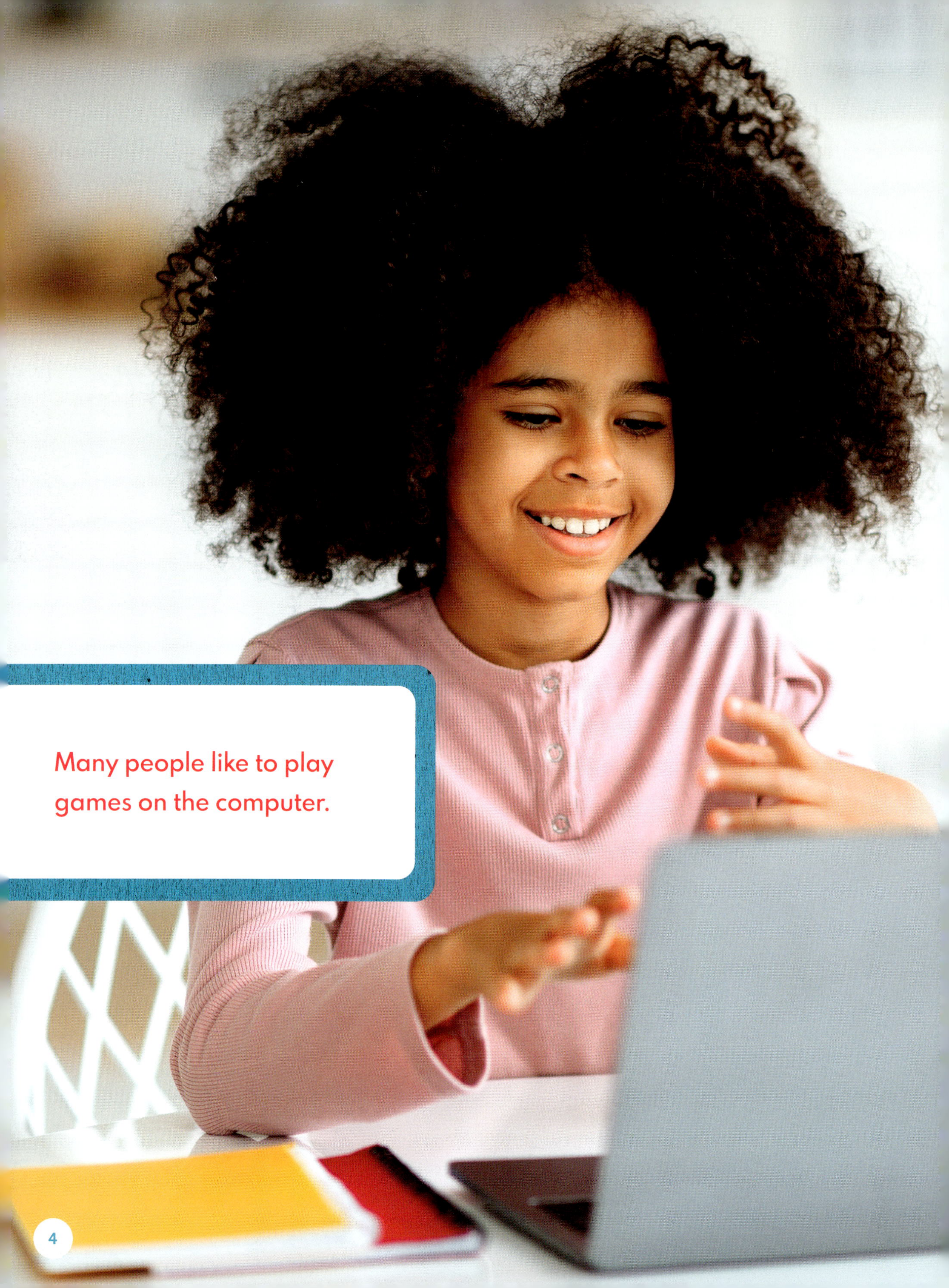

Many people like to play games on the computer.

CHAPTER 1

PASSWORD PROBLEMS

Jess wants to play a game on the computer. Before she starts, the computer asks for her password. She quickly types it in. The screen flashes red. Oops! Jess must have touched a wrong key. The computer tells her to try again. This time she types slowly.

She gets the password right. The game opens on her computer. She's ready to play!

People can think for themselves. They can make choices. But computers are different. They follow the exact instructions people give them. The computer knew Jess had typed the wrong password. This is because a person wrote instructions for the computer. These instructions are called a program. The program told the computer that if the password is correct, then the game can start. If the password is wrong, then the game cannot start.

Programs are written in code. People called programmers write this code. Code can be written in different **programming languages**.

Passwords protect people's digital information.

The computer translates the code. Then it understands the program's instructions.

Coding Conditionals

Conditionals help computers make decisions. People use conditionals in everyday life. One example is *if* and *then* statements. These *if* and *then* statements describe a condition.

For example, a mom might say to her child, "*If* your homework is done, *then* you can play." A brother might tell his sister, "*If* you let me play your new video game, *then* I will let you use my new art set." A dad might say to his child, "*If* you save your money, *then* you can buy more trading cards."

What's in a Name?

Conditionals are part of all programming languages. They are also called conditions or conditional statements. Programmers can use all three names to mean the same thing.

Parents may make rules that kids have to finish their homework before playing.

Programs check whether conditions are true or false in order to make decisions.

Conditionals control what a program does. If something is true, the program runs one piece of code. If something is false, the program runs a different piece of code. The password on Jess's game is an example of a conditional. *If* Jess enters the right password, *then* she can start the game. Programmers can use conditionals in many ways. They are an important tool in writing code.

Further Evidence

Visit the website below. Does it give any new evidence to support Chapter One?

If Statements

abdocorelibrary.com/coding-conditionals

Conditionals allow programs to make decisions.

CHAPTER 2

KINDS OF CONDITIONALS

A computer makes decisions based on conditionals. The conditionals in the code guide the program's choices. They tell the program to do a certain action. The program does the action depending on if the condition is true or false.

A person who comes to a fork in the road must choose which path to take. Conditionals help programs decide what code to run.

Developers can use conditionals to customize apps.

If statements perform the code's instructions if the condition is true. *If* the condition is true, *then* the program runs a block of code. For example, a programmer could create a mobile app. They want the app to greet the user when it is opened. They could use an *if* statement to make the app use a specific greeting in the morning. *If* the user opens the app in the morning, *then* the app will say "Good morning!"

Else statements let a program check multiple conditions.

Else Conditionals

Sometimes the *if* statement is false. Then nothing happens. But programmers can give another instruction. This is called an *if . . . else* statement. The *if . . . else* statement gives two options for running the code. *If* the condition is true, one block of code will run. Or *else*, if the condition is false, different code will run.

The programmer has made their app say "Good morning!" in the morning. But they want the app to say something else if it isn't morning. They can use an *if . . . else* statement to change the greeting. *If* the user opens the app in the morning, *then* the app will say "Good morning!" *Else*, if it isn't morning, the app will say "Hello!"

Sometimes there is more than one *if* statement. *Else if* statements link multiple *if* statements together. Programmers use the *else if* statement to test multiple conditions. The programmer's app already says "Good morning!" in the morning. They could also make it say "Good afternoon!" in the afternoon. The app can test even more conditions if desired.

First, the program will check to see if it's morning. If the condition is true, the program will say "Good morning!" and not check anything else. But if the condition is false, then the program will check to see if it's afternoon. *If* the user opens the app in the morning, *then* the app will say "Good morning!" *Else if* it is not morning but is afternoon, the app will say "Good afternoon!"

Comparison Operators

Comparison operators are symbols. They are often used in conditional statements. They compare two values. These operators check to see if a condition is true or not. Comparison operators are math symbols such as less than (<), greater than (>), or equal to (=).

If, *Else If*, and *Else*

Sample Code

```
if (time < 12:00 p.m.)
  greeting = "Good morning!"
else if (time < 5:00 p.m.)
  greeting = "Good afternoon!"
else
  greeting = "Good evening!"
```

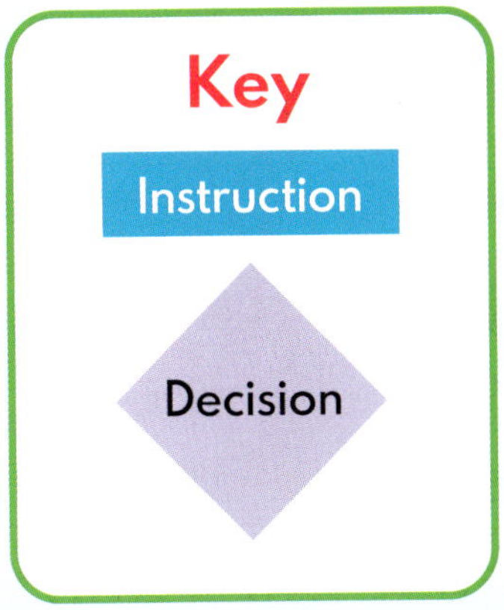

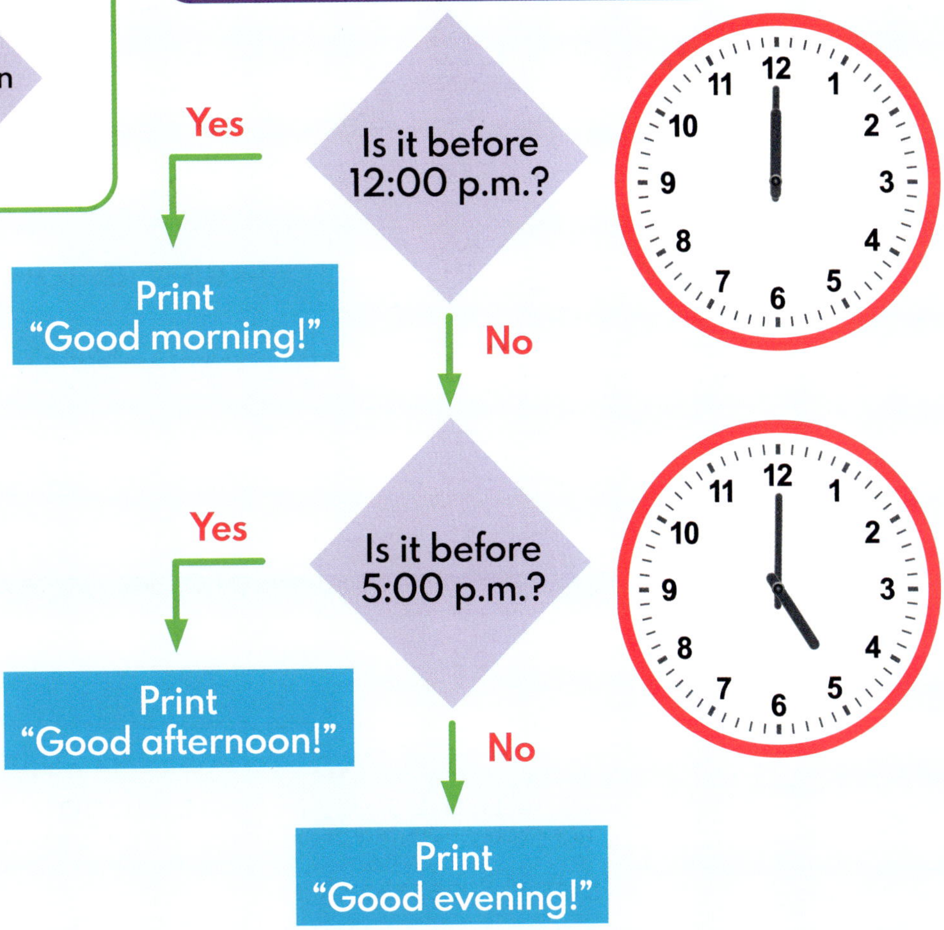

Some programmers create a flowchart before writing code. The flowchart shows how the conditional will work.

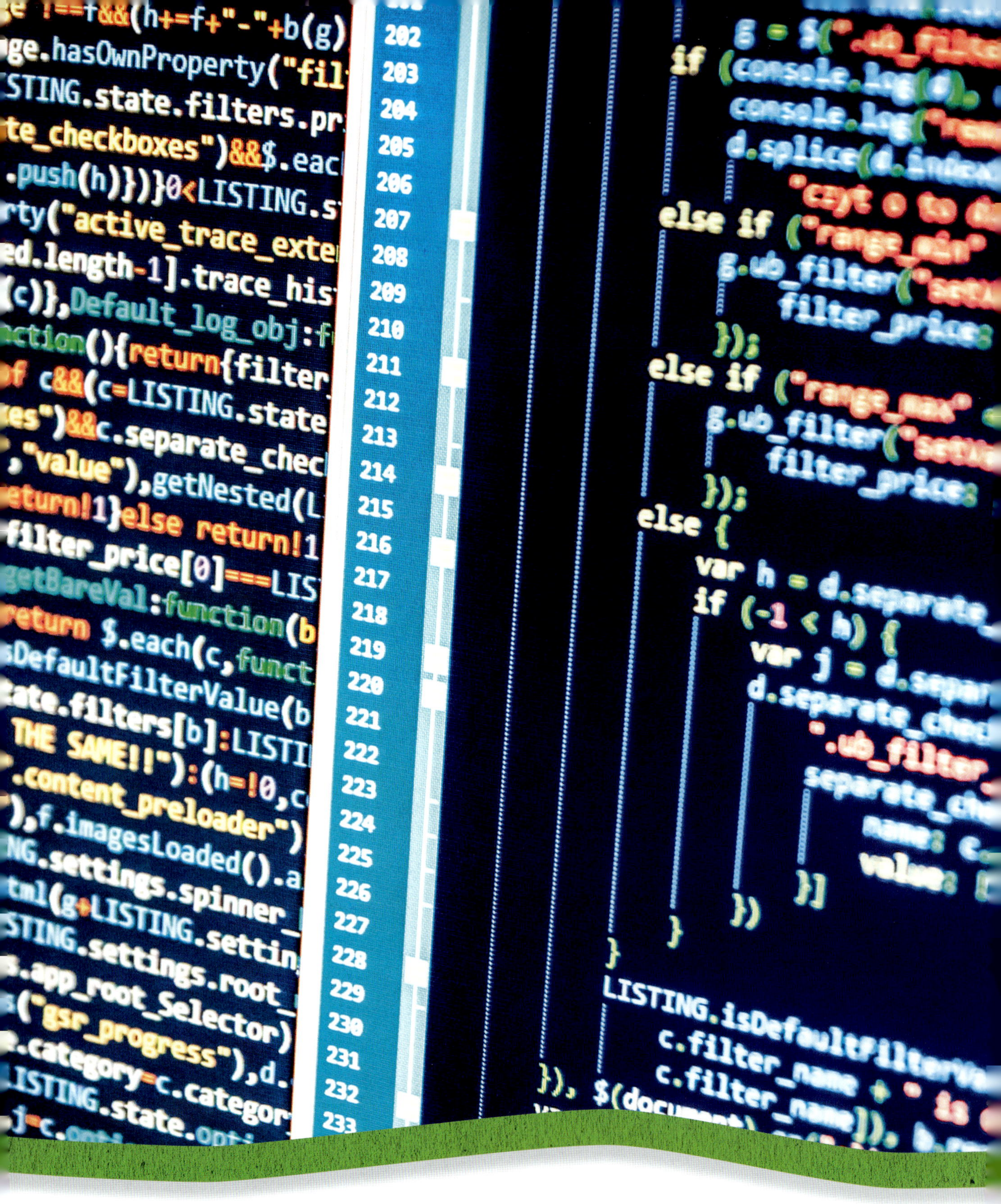

Conditionals are an important part of code.

If, *if . . . else*, and *else if* statements are three basic conditionals. Programmers can use them in many different combinations. This helps their programs make decisions.

Explore Online

Visit the website below. Do the video and activity give any new information about conditionals that wasn't in Chapter Two?

If/Else with Bee

abdocorelibrary.com/coding-conditionals

Making decisions can sometimes be hard for people. But computers can't make decisions without conditionals.

CHAPTER 3

WHY ARE CONDITIONALS IMPORTANT?

Conditionals are a **vital** part of programming. Without conditionals, programs can't make decisions. Programmers would only be able to write very simple programs.

A thermostat controls the temperature of a building.

A simple thermostat might have only three settings. It can blow hot air, cold air, or no air. A person would have to push a button to change the setting. The person would be making the decision, not the program.

But programmers can use conditionals to make a thermostat control its own settings. A person sets the desired temperature. Then the

Details Matter

Syntax is important in spoken languages. Someone may want to say, "We are going to the party later." If he says, "Party are going the to later we," his friends would not understand him. The same is true in code. Code must be written in a specific order. If the syntax is not correct, the computer will not understand its instructions.

Advanced thermostats use conditionals to maintain the correct temperature.

thermostat checks the temperature in the room. If the room is too cold, the thermostat blows hot air. If the room is too hot, the thermostat blows cold air. If the temperature is just right, the thermostat stops blowing air. Conditionals allow the thermostat's **software** to make decisions on its own.

Programmers need to know how to use conditionals in their code.

A Vital Tool

Conditionals are an important tool for programmers. They exist in every programming language. How they are written changes in each language. But they all are an important part of computer code. Without conditionals, programs could not make decisions.

PRIMARY SOURCE

A software engineer described the importance of conditionals:

> Conditional statements are vital in the field of programming and software engineering, [because they] allow a machine to **simulate** the behavior of a person who has the ability to make choices and [act] based on the decision taken.

Source: Abdulhafiz Suleiman Danmaigoro. "What Are Conditional Statements in Programming?" *Educative*, n.d., educative.io. Accessed 8 June 2023.

Comparing Texts

Does this quote support the information in this chapter? Or does it give a different perspective?

CODING DATA

Conditionals allow computer programs to make decisions.

If statements tell a program to run code if a condition is true.

If . . . else statements tell a program to run one piece of code if a condition is true or a different piece if the condition is false.

Else if statements link multiple *if* statements together.

Glossary

programming language
a vocabulary and set of rules for writing computer programs

simulate
to look or act like something else

software
all of the programs that give a computer instructions

syntax
how words are put together to form sentences

vital
extremely important

Online Resources

To learn more about coding conditionals, visit our free resource websites below.

Core Library
CONNECTION
FREE! COMMON CORE MULTIMEDIA RESOURCES

Visit **abdocorelibrary.com** or scan this QR code for free Common Core resources for teachers and students, including vetted activities, multimedia, and booklinks, for deeper subject comprehension.

Booklinks
NONFICTION NETWORK
FREE! ONLINE NONFICTION RESOURCES

Visit **abdobooklinks.com** or scan this QR code for free additional online weblinks for further learning. These links are routinely monitored and updated to provide the most current information available.

Learn More

González, Echo Elise. *Order in Coding.* World Book, 2021.

Kulz, George Anthony. *Coding Loops.* Abdo, 2024.

Lynn, Randy. *Coding Concepts for Kids: Learn to Code without a Computer.* Rockridge Press, 2020.

Index

About the Author

Donna B. McKinney is the author of more than 20 nonfiction books for kids on topics ranging from science and technology to history to sports. Before she wrote for children, she spent many years writing about science and technology topics at the US Naval Research Laboratory in Washington, DC. She lives in North Carolina.